**FAYSSAL Cheriet**
**Hosni Taleb**

# Landslide analysis with the finite element method

**FAYSSAL Cheriet**
Hosni Taleb

# Landslide analysis with the finite element method

**ScienciaScripts**

**Imprint**
Any brand names and product names mentioned in this book are subject to trademark, brand or patent protection and are trademarks or registered trademarks of their respective holders. The use of brand names, product names, common names, trade names, product descriptions etc. even without a particular marking in this work is in no way to be construed to mean that such names may be regarded as unrestricted in respect of trademark and brand protection legislation and could thus be used by anyone.

Cover image: www.ingimage.com

This book is a translation from the original published under ISBN 978-620-2-28395-3.

Publisher:
Sciencia Scripts
is a trademark of
Dodo Books Indian Ocean Ltd. and OmniScriptum S.R.L publishing group

120 High Road, East Finchley, London, N2 9ED, United Kingdom
Str. Armeneasca 28/1, office 1, Chisinau MD-2012, Republic of Moldova, Europe
Printed at: see last page
ISBN: 978-620-5-94535-3

# TABLE OF CONTENTS

# SUMMARY

This book is composed of two phases. The first phase deals with the generalities of landslides, their classifications and the different methods of landslide analysis. The second phase deals with a comparative study of landslides (Case of Parc Dounia Alger, ALGERIA), this comparison was made by two softwares (PLAXIS and GEOSLOP) with a different calculation basis. The first one is based on the finite element method with the technique of mechanical parameters reduction (c-phi reduction) until failure. The second is based on limit equilibrium methods.

**Key words**

Landslide, Classification, Analysis methods, Modelling, c-phi reduction, PLAXIS, GEOSLOPE.

# GENERAL INTRODUCTION

Among the computational methods used to analyse a slope stability problem are classical methods (limit equilibrium calculation by the slice method) and numerical methods (finite difference methods, finite element methods). The latter is popular because of its advantages (studying media with homogeneous or heterogeneous mechanical characteristics, taking into account various rheological behaviours, finding the automatic sliding surface and the lowest safety factor).

This book consists of two parts. The first part deals with generalities on landslides such as classification, causes and different methods of landslide analysis. The second part is a landslide analysis (case of Parc Dounia in Algiers, Algeria) with two different basic methods (finite element and limit equilibrium), one is based on the c-phi reduction method (reduction of the mechanical resistance parameters) using PLAXIS 2D code version 8.2 and the other is based on the ordinary slice method using GEOSTUDIO code (SLOPE/W module).

**PART I**

**GENERAL INFORMATION ON LANDSLIDES**

In this chapter, we present a generalization on landslides. The first part of this chapter will be devoted to a synthesis of knowledge on the different classifications of landslides. We present a general overview of the kinematics of mass movements, the landslides observed throughout the world and their causes. The second part is devoted to the different calculation methods.

## I.1 Introduction

Slope stability problems are one of the phenomena often considered as natural hazards triggered and reactivated by the force of nature (soil properties, slope angle, presence of water, ...). However, human action is often preponderant in this type of hazard and is one of the most common triggers of instability. And other times with the intervention of the two causes (nature and human). These instabilities cause significant damage: they affect natural slopes as well as artificially created slopes, and pose a threat to infrastructure or inhabited areas. Since the construction of many infrastructures requires the installation of slopes, the analysis of the stability of these slopes includes, in addition to the knowledge of the site, the choice of the mechanical and hydraulic characteristics of the soil. The estimation of this stability with respect to the risk of failure is one of the important problems in geotechnics, especially in the field of limited or unknown data. In this section we will present a general overview of landslides, on the one hand classification; kinematics; landslides observed throughout the world, and their causes and on the other hand the calculation of their methods.

## I.2 Classification of landslides

There are many classifications aimed at describing and determining the different processes of slope destabilisation. As early as 1846, Collin [1], in his work "Recherches expérimentales sur les glissements spontanés des terrains argileux" (Experimental research on spontaneous landslides in clay soils) differentiated two types of movements:

- Ground movements (deep slides);
- Superficial movements.

Many authors have presented several classifications of ground movements, using criteria such as: nature of the material, kinematics of the movement, velocity of the movement, morphology and type of failure surface, cause of failure, age of failure, etc. The classification of Nemcok et al [1] is based on the geomechanical character of the slope movements and their velocity, four (04) processes are distinguished: creep; slip; flow and collapse.

According to Varnes' classification, [1] the types of slope movements are: Landslides, Slides, Lateral extension, Flow, Complexes, Velocity scale. The Varnes classification also proposes a velocity scale to characterise the degree of activity of the different movements (Figure 1.1).

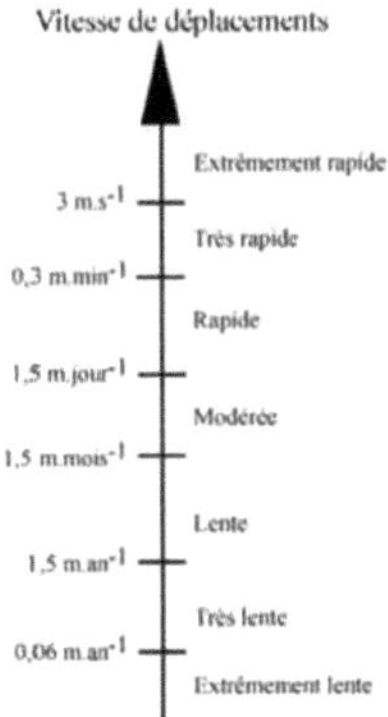


Figure. 1.1 Degree of activity as a function of travel speeds

In the same context of kinematic classifications. The degree of activity of the different movements can also be classified through a scale of displacement speed of the mass involved. The intensity of the landslide then varies in space from upstream to downstream along the axis of movement of the landslide. These parameters can be expressed quantitatively, e.g. by using a spatial distribution function, or by applying relative qualitative intensity rankings such as slow, moderate and fast, or low, moderate and high. The following table by Cruden and Varnes in 1996 attempts to establish a landslide intensity scale taking into account the speed of movement of the mass involved.

Table 1.1 Classification of landslides according to their displacement speed Vd (Cruden and Varnes, 1996) [1,2].

| Speed | Description |
|---|---|
| $Vd$ < 16 mm/year | Extremely slow Very slow |
| 16 mm/year< $Vd$ < 1.6 m/year | Slow Moderate Fast Very fast Extremely fast |
| 1.6 m/year< $Vd$ < 13 m/month | |
| 13 m/month < $Vd$ < 1.8 m/h | |
| 1.8m/h<yd < 3 m/min 3 m/min < $Vd$ < 5 m/s | |
| Vd >5 m/s | |

Meunier (1991) [1], classified the different types of slope destabilisation processes according to two main poles: one represented by water and the other by solid materials (Figure 1.2).

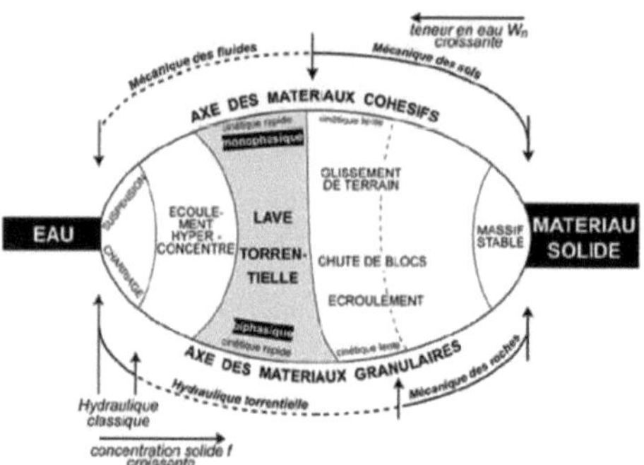


Figure. 1.2 Classification of ground movements according to Meunier (1991)

Starting from the water pole, it can be determined from which solid load the transported solid material changes the behaviour law of the mixture and influences the flow height. In other words, from a certain solid concentration limit, the classical way of proceeding (water equations, river hydraulics) is no longer possible. By increasing the solid load within the flow, we progressively move to fluvial hydraulics (scouring, suspension), then to hyper-concentrated flow and finally to debris flows. If we now start from the solid material pole, we study the water content at which a movement can be triggered. The scientific problem consists in studying the rupture and the stability conditions of the ground. The aim is to determine the (relatively unknown) limits at which a flow or a ground movement (landslide, rockfall, collapse) can be transformed into a debris flow. Landslides in soils and loose rocks occur in many forms (Table 1.2; planar or circular landslide, lateral displacement, mudflow, debris flow, creep), over varying areas (a few $m^2$ to several $km^2$ ; Table 1.3).

Table 1.2 Typological classification of landslides (adapted from Varnes, 1978) [3]

| Type of slip | | Type of soil or loose rock | |
|---|---|---|---|
| | | Rough | End |
| Slippage | Rotational | | |
| | Translational | Debris slide | Mud slide |
| Lateral displacement | | Displacement of debris | Moving the boat |
| Casting | | Debris flow | Mudflow Reptation |
| Complex | | Combination of at least two mechanisn | |

Landslides are displacements of materials (loose or indurated fine rock, natural or anthropogenic soil) along a basal shear surface.

Table 1.3 Classification of landslides by area (Cornforth, 2005) [4]

| Area (m )$^2$ | Description |
|---|---|
| < 200 | Very small |
| 200 < s < 2000 | Small |
| 2000 < s < 20000 | Medium |
| 20000 < s < 200000 | Big |
| 200000 < s < 2000000 | Very large |
| s > 2000000 | extensive |

## 1.3  Kinematics of mass movements

Each type of ground movement has different types of temporal evolution. Flageollet (1996) [1], notes that some of these movements can stop immediately and definitively after the triggering (boulder falls, collapse), while others experience a cyclical activity (torrential flow).

For landslides, Vaunat and Leroueil (2002) [1], distinguish four (04) stages of temporal activity in terms of kinematics (Figure 1.3 (a)):

- The pre-fracture stage, when the initial mass is still continuous and characterised by the onset of progressive failure or creep;
- The failure stage, characterised by the formation of a continuous shear surface over the entire unstable mass;
- The post-breakdown stage writes the behaviour of the mass from the propagation phase to the stop (deposition) phase;
- The reactivation stage when the sliding mass flows along pre-existing surfaces. Reactivation is either occasional and sudden or continuous and characterised by a seasonal rhythm of movement speeds.

a) Temporal dynamics (stages of activity) of a landslide

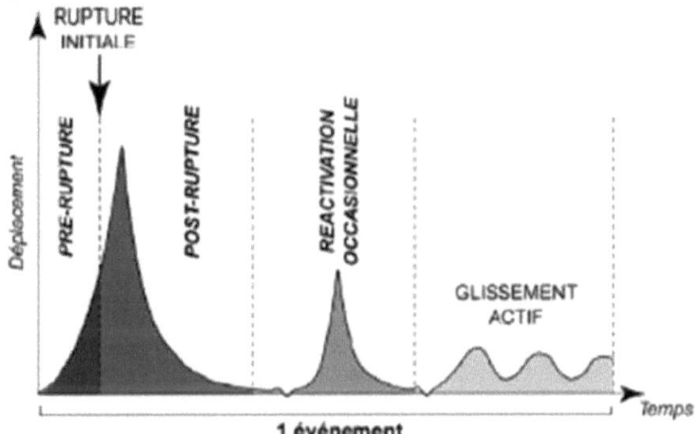


b) Temporal dynamics (stages of activity) of debris flows

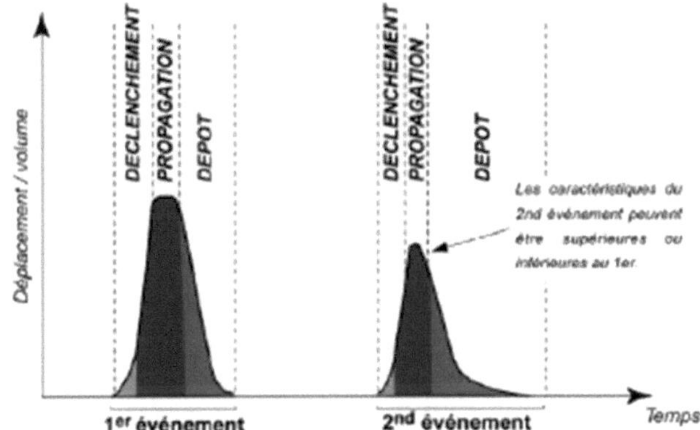

Figure. 1.3 Comparison between the temporal dynamics of (a) landslides and (b)
debris flows
(after Vaunat and Leroueil, 2002)

## 1.4  Landslides around the world

As a statistic, the costs related to landslides (follow, controlled,...) between 1 - 5 USD billion, per year for the following countries: USA, India, Japan and Italy [6]. In Canada the cost is about 70 million dollars [7]. The major damage caused (destruction of infrastructure and devastation of villages) results in a high cost rate [6,8]. They make daily life difficult as a result (cutting of roads, distriction of pipes, reduction of agricultural and forest land).

### 1.4.1  In Algeria

The geomorphology of northern Algeria is essentially characterised by mountains with steep and abrupt slopes. This phenomenon has been observed in several northern wilayas of the country: Algiers, Béjaia, Constantine, Mila, Médéa, Tizi-Ouzou.... etc. These regions are experiencing activities of this hazard. Taking some examples.

#### a) Béjaia

As shown in Fig. 1.4, the land affected by the landslide has a steep slope of more than 60° and is prone to landslide, due to the high rainfall affecting the Bejaia region. Concerns a road with a mixed cut and fill profile in a mountainous area. The area extends over a length of 80 m. The geotechnical investigations revealed, from the surface, sandstone scree, sandstone banks and alternating schistose marls [9].

Figure. 1.4 Longitudinal cracks in the road affected by the slip

#### b) Constantine

The region of Constantine is located in a complex geomorphological environment and on one of the most important seismic axes of Eastern Algeria. The instability of the ground in Constantine began to appear for the first time at the beginning of the 20th century[éme] during and especially after the increased development of the city and the construction of engineering works such as the 'Sidi Rachad' bridge in the eastern part. Since then, the city and its surroundings have not ceased to experience degradation (landslides and others), which have affected the best known districts: the most recent event which affected the Aouinet El-Foul district, dates from 1972, the year during which major disorders had affected the entire district of KITOUNI. After this date, and particularly since a few years, the Constantine region has been subjected to changes in the plans for hazardous development on land that can no longer support overloads (figure 1.5) [10].

Figure. 1.5 Consequence of the Constantine landslides

### 1.4.2 Across the world

A few shots of five cases of slope desloping. The examples chosen are the following: Philippines, Puerto Rico, Hiroshima (Japan), Oso Washington (USA) and Shenzhen (China) [11].

Figure. 1.6 Puerto Rico landslide, 1985 (Randall Jibson, U.S. Geological Survey)

In 1985, a large landslide occurred in Mameyes, Puerto Rico (Figure 1.6) following a storm that produced extremely heavy rainfall. This calamity devastated about 120 houses and caused the loss of 129 lives.

Figure. 1.7 Philippines landslide, February 2006 (University of Tokyo, Geotechnical Group)

Photo in Figure 1.7 (University of Tokyo Geotechnical Group) shows the landslide (avalanche) that buried the village of Guinsaugon, southern Leyte, Philippines, in February 2006, following a period of heavy rain.

Figure. 1.8 Hiroshima landslide Japan, 20 August 2014 (Reuters Kyodo)

Figure. 1.8 (Reuters Kyodo) shows a shot of a landslide in Hiroshima, Japan that occurred on 20 August 2014. The mud swept through a residential area where most of the buildings were recently constructed.

(a) Air Support Unit (b) County Sheriff's Office

Figure. 1.9 'Oso' landslide, Washington, USA, 22 March 2014

22 March 2014, a large landslide occurred in Oso, Washington USA, figure. 1.9 (a) (The air support unit), shows the spread or scale of the affected area, figure. 1.9 (b). (County Sheriff's Office) the extent of the destruction following a mud and debris flow.

Figure. 1.10 Shenzhen landslide, China, 21 December 2015 (Xinhua)

Monday 21 December 2015. The landslide occurred in the city of Shenzhen near Hong Kong, this city is among the largest in China. In its statement, the Ministry of Land and Resources of China, on its website "Because the mound was very large, and the angle of the slope was too steep, which led to the loss of stability. Another landslide in Xinhua caused a natural gas pipeline to explode. As a result of this disaster, seven people were pulled from the rubble with varying degrees of injury, and 91 people were reported missing, along with the destruction of 33 buildings. Shown in Fig. 1.10 (a), and figure. 1.10 (b).

Figure. 1.11 Landslide in Sierra Leone. 14 August 2017

More recently, on 14 August 2017, a landslide occurred on the outskirts of Freetown, the capital of Sierra Leone. The landslide was caused by a night of heavy rain. The latest death toll is 312, according to the Red Cross.

Table 1.4 shows landslides around the world with a brief description.

Table 1.4 Landslides in the world with a brief description

| Location (country) | Date | Deaths | Note |
|---|---|---|---|
| Mont Granier (France) [12] | 25, November 1248 | 1,000+ | Five villages destroyed. |
| Kelud volcano (Indonesia) [13]. | 1919 | 5,160 | Drainage of the lake |
| Gansu province (China) [13]. | 16, December 1920 | 180,000 | flows affected an area of 50,000 $km^2$ |
| Kwansai, (Japan), [12] | 5 July, 1938 | at least 1000 people | 130,000 houses damaged or destroyed |
| Khait (Tajikistan), [13] | 1949 | 12,000 | Rock separated by an earthquake. |
| Longarone, Belluno (Italy), [13] | 9 October, 1963 | 2,000 | Rock avalanche on an artificial water reservoir |
| Hurricane Camille, central Virginia (USA), [14] | 1969 | 150 | Killed by broken bones and other blunt force injuries |
| Mameyes (Puerto Rico), [15] | 1985 | at least 129 people | 120 houses destroyed |
| Villa Tina, Medellin (Colombia), [13] | 27, September 1987 | 217 | shallow slide in lateritic residual soils |
| Manjil (Iran), [16] | 20 June 1990 | More than 200 people | these landslides, caused by the earthquake. Destroyed or buried several villages, blocked several roads. |
| Casita, Nicaragua (Central America), [17] | 1998 | More than 2500 people | Triggered by heavy rains during Hurricane Mitch. |
| Pizzo d'Alvono, Campania (Italy), [18] | 5-6 May 1998 | 160 | 150 buildings are totally destroyed and over 500 points of serious or partial damage |
| Taiwan, [19] | 21 September 1999 | More than 2200 people | Caused by (The Chi-chi) earthquake of magnitude 7.3 |
| Santa Tecla, El Salvador, (Central America), [20] | 2001 | 500 | Landslide, occurred during a major earthquake, involving about 180,000 |

| | | | $m^3$ of volcanic deposits, had a total drawdown of about 800 m. |
|---|---|---|---|
| La Conchita, California (USA), [13] | 10, January, 2005 | 10 | A debris flow mobilised previous landslide deposits due to heavy rains, destroying or damaging about 30 houses in the community |
| Guinsaugon, St. Bernard, Southern Leyte, (Philippines), ' [12] | 17, February, 2006 | 1126 | Debris avalanche triggered by ten days of heavy rain |
| Chittagong, (Bangladesh), [12] | 11, June, 2007 | 123 | Series of landslides caused by monsoon rains and illegal hills |
| Cairo, (Egypt), [12] | 6 September, 2008 | 119 | Rock falls from cliffs, individual rocks up to 70 tonnes |
| Kedamath, Uttarakhand, (India), [12] | 16, June, 2013 | 5,700 | Caused by high rainfall intensity |
| Oso, Washington (USA), [12] | 22 March, 2014 | 41 Confirmed missing | The flow of the landslide was extreme because of the extraordinary performance of the mud and debris. |
| Ab Barak, Badakhshan (Afghanistan), [12] | 2 May, 2014 | 2,000 | The landslides were caused by heavy rainfall |
| Hiroshima (Japan), [21] | 20 August, 2014 | 74 | Caused by heavy rains |

## 1.5  Causes of landslides

Many factors make slopes unstable. Some are natural, some are human and some are a combination of both. In general, it is unlikely that there is an isolated reason that acts alone and causes slope instability. Here we will present natural and human causes of landslides.

### 1.5.1  Natural causes

This category includes various natural causes that can occur individually or in combination. Among the main causes: slope configuration, geological aspect, mechanical properties and soil class, water, erosion, earthquake, weather, etc. The configuration of the slope is related to geometric data such as slope, declivity and orientation of layers in relation to the slope. Geology is involved in the relief of geomorphology and stratification. Mechanical properties are fundamental. Indeed, they affect the stability of the slope and determine the interaction of the layer interfaces. Variation due to hydraulic conditions as we will see below among the direct causes of landslides. Water plays an important role as one of the main factors of landslides. The existence of water near the slope occurs with many ways as precipitation, rivers, dams, lakes, groundwater, etc... The rate of pore pressure changes can be slow or fast depending on the permeability of the soil. In addition, water increases the weight of the soil unit and can lead to a reduction in soil shear forces as the pore pressure increases, [22]. The other natural factor is seismic activity. Many landslides have occurred during earthquakes. The earthquake causes soil expansion and lead to rapid water infiltration and reduction of shear strength, [23,24]. Various weather conditions also influence the slopes. Such as watercourses, chemicals and mechanical actions. Mechanical weathering is a consequence of the following factors, freeze-thaw cycles, and temperature change, erosion by streams, wind, trees and vegetation roots. These causes break down the rocks into smaller fragments and reduce the shear strength of the soil due to the power developed by the physical forces, [24]. The effect of chemical weathering can be influenced after a few days to several years, [25]. Chemical changes due to the following factors: acids in the river water, in the air, and in the rain. In addition to the properties of the slope material, chemical weathering depends on other factors such as climate, geological details, and drainage. The actions of these mineral factors degrade the soil to new composites. Bjerrum (1967) discussed chemical weathering in consolidated clays in his state-of-the-art paper on progressive slope failure in more plastic consolidated clays and clay-shales [26]. There are many other causes in this category such as volcanic eruptions that require special considerations and monitoring. In the investigation of slope stability, the most important task is to know the causes and understand the evolution of the failure mechanism.

## 1.5.2 Human causes

There are several human factors contributing to the occurrence of landslides. Population growth on new land with the creation of cities, this urbanisation and development magnifying the danger of many types of disasters, slide causes. Human activities can result in a change in the morphology of the slope or soil shape: illustrations include slope steepening or weakening of the footing, creating cracks during excavation or deforestation. Other causes associated with additional loading: there are many types of loading can be short-term, such as passing vehicles, vibration from machinery; or can be permanent, for example the construction of a building. Unstable slopes can also occur through causes due to other human activities related to water, such as leaking pipes, creation or drainage of reservoirs, watering of lawns and disturbing or variable drainage patterns [11].

## 1.6  Slip calculation

Two types of calculations can be made.

### 1.6.1  Pre-slip calculation (a priori study)

The most critical geometry and the most unfavourable surface in this case are not known a priori. The objective of the calculation is to determine which of the infinite number of possible failure surfaces will be the most critical. The calculation will therefore consist of testing as many surfaces as possible and finding the most unfavourable surface by "trial and error". Each surface tested will be the subject of a stability calculation which will provide, in general, the value of a safety coefficient "Fs". Fs is the safety coefficient of the slope with respect to failure on the surface under consideration. The safety coefficient of the site will be the lowest of the Fs values obtained. The area with the lowest safety coefficient is the most likely failure area.

### 1.6.2  Post-slip calculation (hindcast) [1].

In this case, it is a matter of understanding and analysing the slide (in particular to avoid the recurrence of other slides under the same conditions). The aim is to improve the situation in such a way as to achieve acceptable safety. In this case, the geometry of the failure surface is known (at least partially) and, since there has been a failure, this means that the land has reached its failure limit.

Stability analysis is an important part of the design of embankments, slopes, excavations, dams, etc.

## 1.7  The choice of the type of calculation method [1]

Another important choice, which depends on the means that can be used, is to

between a method that models the entire soil mass (finite element method) and

a kinematic method, defining a failure surface for example (limit equilibrium method) However, with the possibilities of analysing a large number of

In the case of a potential failure curve, the two approaches are similar. In the case of a

A method based on a previously defined curve will be repeated many times to obtain a similar result. This choice must be made by examining the means available, the overall behaviour of the slope, but also by ensuring that the calculation parameters corresponding to the model can be obtained.

The overall behaviour of the slope corresponds to four mechanisms that result in differently distributed soil displacements [27]:

- **Pre-breakdown**, where the behaviour of the soil is elasto-visco-plastic and where the mass is a continuous medium, without any discontinuity zone, the deformations are almost homogeneous;

- **Fracture**, where one part of the mass moves relative to the other, the soil model is elasto-plastic or even rigid-plastic;

- **Post-breakup**, where one part of the soil moves over the other, as a viscous flow and with appreciable speed;

- **Reaction**, when the part of the ground that has already slipped and stabilised, resumes movement on a surface, following a rigid-plastic behaviour;

The distinction between these four mechanisms is fundamental for a reliable study of slopes, and this will of course influence the choice of a calculation method. It helps to choose between the types of methods. Generally speaking, there are two main coupled tasks in slope stability analysis:

1. The calculation of the safety factor ;
2. Location of the critical slip surface (the slip line).

## I.8 Methods of analysis

There are several methods for assessing slope stability. Historically, the analysis is performed by analytical methods such as limit equilibrium and limit analysis. With the development of computers, numerical methods have experienced a tremendous rise. Their robustness consists in the ability to conduct coupled analyses in multi-physics fields, faster and with sufficient accuracy. In the slope stability problem, numerical analysis automatically searches for the critical failure surface and the lowest safety factor [28]. Numerical methods are quite varied and can be classified into frameworks of continuous and discontinuous methods, hybrid methods, optimisation methods, artificial intelligence methods, etc. The following sections provide a brief description of some of the methods.

### 1.8.1 Limit equilibrium method

For many years the limit equilibrium method (LEM) has been the most popular method to analyse the stability of slopes. Until now, it is still used for simplicity, as it does not need many parameters. The usual assumptions are the theory of plasticity, the Mohr-Coulomb failure criterion applied over the whole sliding surface, the single value of the safety factor (Fs) over the whole sliding surface, the sliding surface is known or experienced and the failing mass is rigid. There are many methods based on this technique. Some limit equilibrium methods can be obtained by manual calculation, such as the infinite slope analysis [29]. Therefore, abacus methods were created to avoid repeating the calculation [30]. These graphical solutions provide a quick check of the analysis results, and are generally useful for preliminary analysis. The most important limitation in the use of design abacuses is that most cases are ideal, which does not correspond in practice, such as homogeneous soil conditions [31]. The abacuses were developed using the following general assumptions: two-dimensional limit equilibrium analysis, circular sliding surfaces and homogeneous slopes. However, currently computer software is mainly used to perform slope stability analyses. Among the numerical methods based on limit equilibrium, the slice method is widely used. The practical calculation of the minimum Fs is based on the circular sliding surface [32]. The method consists of dividing the slope into finite vertical slices bounded by the free surface and the failure circle (Figure 1.12). The definition of the safety factor (Fs) is the same for all variants of this method:

$$FS = \frac{Actions\ stabilisatrices}{Actions\ Motrices} \qquad (1)$$

The safety factor is calculated for each individual slice and then for the whole body (ABCD). In this method, the mass of soil (ABCD) above a test failure surface (AC) is a circular arc of centre O and radius R. The mass (ABCD) is divided into n slices by vertical planes (Figure, 1.12). As well as inclined slices have also been used by different researchers [33]. In general, the differences between the various cutting methods are not significant, and the vertical cut is favoured by most engineers. Figure 1.12 shows one of the many circles for which the safety factor must be determined, where b is the width of the slice, a is the slope of the base of the slice and h is the height measured on the centre line. The forces acting on a slice are : $W = bhy$, is the total weight of the slice. $N = al$, is the total normal force on the base. This force has two components, the water force, $U = ul$, where u, is the pore pressure of water at the centre of the base, and $l$ is the length of the breaking surface for the slice. The effective normal force $N$ (equal to $al$) and $T = z_m\ l$ is the shear force on the base. $E1$ and $E2$, are the total normal forces on the sides. $X1$ and $X2$ *are* the shear forces on the sides.

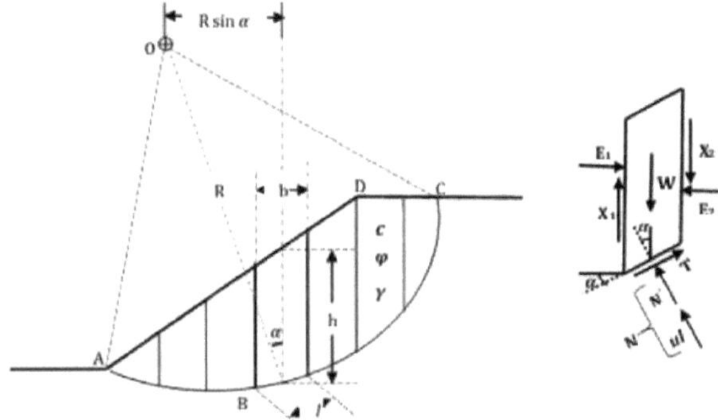


Figure. 1.12 Slicing method

The difference between the variants of slice methods represented in the global equilibrium conditions is the assumptions about the forces between the slices and the shape of the sliding surface. Here we will give a brief discussion of some of the methods. The method of Fellenius, (1936) [34], known as the Ordinary Slice Method (OSM), is one of the simplest procedures. This method neglects all forces between slices, assumes the normal force on the base of each slice, and applied only to the circular fracture surface, and the analyses are extremely inaccurate for the effective stress of slopes with high pore pressure. The safety factor in terms of effective stress is given by :

$$FS = \frac{c'L_a + \tan\varphi'\ \Sigma(W\cos\alpha - ul)}{\Sigma W \sin\alpha} \qquad (2)$$

15

Where $L_a$ is the arc length AC. Bishop's simplified method, which has been widely used, is a fairly accurate method used only for circular sliding surfaces, developed by Bishop, (1955) [35], does not satisfy the horizontal balance of force, but does verify the vertical balance and the overall balance of moment.

$$X_1 - X_2 = 0 \tag{3}$$

The formulation of the safety factor requires an iterative procedure and is given by

$$FS = \frac{1}{\sum W \sin \alpha} \sum \left[ \{c'b + (W - ub)\tan \varphi'\} \frac{\sec \alpha}{1 + (\tan \alpha \, \tan \varphi'/FS)} \right] \tag{4}$$

Other slice methods for the analysis of non-circular sliding surfaces are more difficult because they contain more variables. In this context, the method of Morgenstern and Price, (1965) [36], developed to analyse the circular and non-circular sliding surface, is mentioned. The method allows the inclinations of the lateral forces to be varied and calculated in the solution procedure, it satisfies all equilibrium conditions, so it is an accurate method. In 1967, Spencer developed the method known by Spencer, [37]. It is an accurate method because it satisfies all equilibrium conditions, and was formulated to analyse any shape of sliding surface. The lateral forces are assumed to be parallel. The simplified Janbu method was developed by Janbu, [38], it satisfies all equilibrium conditions and assumes that the lateral forces are horizontal (same for all slices). Among the methods that satisfy the total equilibrium conditions, the simplified Janbu method generally gives the lowest safety factor. Other methods are the Lowe and Karafiath method, [39], the modified Swedish method developed by the US Army Corps of Engineers, [40], and the Sarma method, [41], etc. Various slice methods have received considerable attention in the literature, and have been well summarised and reviewed in numerous articles, e.g. Fredlund and Krahn, [42], Transportation Research Board, [43], Duncan, [28], and Abramson et al, [31].

a) Limitations of limit equilibrium methods:

1. The complication of all these limit equilibrium methods is that they are based on the assumption that the mass likely to slip is divided into slices and this implies additional assumptions about the forces between slices and hence about equilibrium. For all methods that satisfy all equilibrium conditions, FREDLUND et al. show that the assumptions made have no significant effect on the safety coefficient; however, in methods that satisfy only the force equilibrium, the safety coefficient is significantly affected by the assumed inclination of the inter-slice forces, [28] which is why these methods are less used compared to others that satisfy all equilibrium conditions.
2. In stability analysis by limit equilibrium methods, the behaviour of the soil is assumed to be rigid and perfectly plastic, so they do not give any information on displacements.
3. The safety coefficient $Fs$ is assumed to be the same at each point of the slide plane. However, we see in Fig. 1.13 that the ultimate shear strength is not necessarily mobilised simultaneously along the slide surface.
4. For complex geometries, there may be a local minimum that remains undetected and complex (non-circular) fracture surfaces may be difficult to detect.

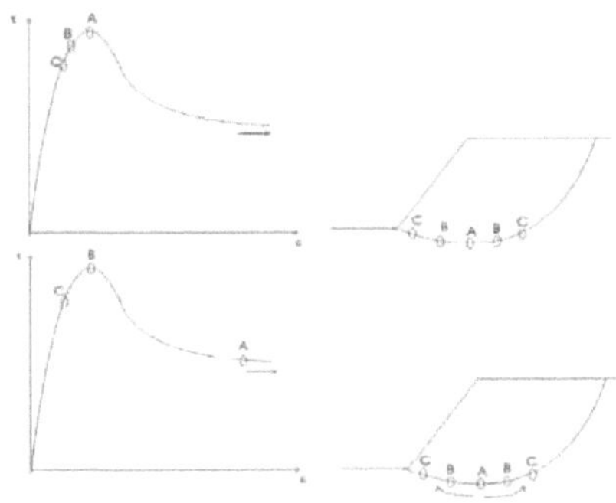

Figure. 1.13 Mobilisation of shear strength along a slip plane, [21]

## 1.8.2 The finite difference method

The Finite Difference Method (FDM) is one of the oldest methods for modelling continuous media. A numerical application of FDM underwent a period of intense development during the 1950s and 1960s [44]. The evolution of FSK techniques was stimulated by the emergence of computers that provided a practical framework for solving more complex problems. SDM consists of replacing derivatives by finite differences where continuity derivatives become discrete differences. In the theory of elasticity, the differential equations are transformed into a linear algebraic system and the solution can be obtained numerically. The discretisation and solution procedure is relatively simple in this method, which is gradually becoming popular. Antonio Bobet, gives a brief description of numerical methods in geomechanics [45]. In recent years, the majority of published geotechnical problems have used finite difference methods, through the Fast Lagrangian Analysis of Continua (FLAC) program [46]. In the study by Dawson et al, a detailed discussion and application of FDM can be found, [47]. Other research on the stability of reinforced slopes has used the FDM. Wei and Cheng, gave a detailed report of nailed slopes under different conditions [48]. Xinpo et al, used pile slope stabilisation analyses [49]. Iman et al, used MDF to study geocell reinforced slopes [50].

## 1.8.3 The finite element method

The finite element method (FEM) is widely used for solid mechanics and structural analysis, [51-54]. FEA has become a more flexible tool, with pre and post processing and high realism associated with complicated behaviour laws, difficult geometry, loading types and coupled multiphysics problems. Generally, there are two most widely used FEM methods for slope stability analysis. The first, Gravity Augmentation Method (GIM), where the gravity forces such a weight gradually increases until the instability of that slope. To give more reliable results, the method is used to study the construction of embankments, [55]. Li, implemented the GIM in a numerical code "Realistic Failure Process Analysis" (RFPA), [56,57]. The second is the c-phi reduction method, which is the most widely used method in FEM and SFM programs for slope stability analysis. The c-phi reduction method gradually decreases the strength parameters (c, $\varphi$) of the slope until it breaks, this method was used by Zienkiewicz et al [58]. The first source code used for slope stability analysis with this technique was published by Smith and Griffiths, [59]. Subsequently, it has been applied by others, e.g. Naylor, [60], Donald and Giam, [61], Ugai, [62], Matsui and San, [63], Ugai and Leshchinsky, [64], Song, [65], Dawson et al. [47], Griffiths and Lane, [66], Hammah et al. [67], Zheng et al. [68]. This method has been reviewed by Duncan, [28], Griffiths and Lane, [66] and Taleb and Berga [69]. The shear strength reduction method is popular for slope stability analysis with Mohr-Coulomb failure criterion, but it is extended to other criteria, such as the Hoek-Brown criterion, on paper by Hammah et al. [70], Taleb and Berga, [71], they are used the generalized Hoek-Brown criterion, for the evaluation of rock slopes. The Hoek-Brown efficiency function, [72] is given by:

$$F = -(\sigma_1 - \sigma_3) - \sigma_{ci}\left(-m_b\frac{\sigma_3}{\sigma_{ci}} + s\right)^a \tag{5}$$

$$m_b = m_i\exp\left(\frac{GSI - 100}{28 - 14D}\right) \tag{6}$$

$$a = \frac{1}{2} + \frac{1}{6}\left(e^{\frac{-GSI}{15}} - e^{\frac{-20}{3}}\right) \tag{7}$$

$$s = \exp\left(\frac{GSI - 100}{9 - 3D}\right) \tag{8}$$

Where, $\sigma_1$ and $\sigma_3$ are the major and minor principal stresses respectively, and GSI: Geological Strength Index, $\sigma_{ci}$ the uniaxial compressive strength of intact rock [kPa]. $m_i$ the intact rock parameter. D, the disturbance factor. GIM and c-ph reduction are used for two tasks coupled with slope stability analysis: on the one hand calculating the minimum safety factor; on the other hand finding a critical slip surface related to this safety factor. Among the advantages of these GIM and c-ph reduction methods, the critical failure surface is found automatically [48]. The safety factor with GIM is the ratio of the gravitational acceleration in the failure time to the actual gravitational acceleration, and is defined according to equation :

$$FS = \frac{g_0^{trail}}{g_0} \tag{9}$$

$g_0^{trail}$ : Test gravitational acceleration (m/s$^2$ ); and $g_0$ initial gravitational acceleration (m /s$^2$ ). With Mohr-Coulomb, the safety factor is the ratio of the actual strength parameters to the critical strength parameters. The Mohr-Coulomb failure criterion can be written as the following equation:

$$\tau = c + \sigma_n \tan\varphi \tag{10}$$

Where, $\tau$ is the shear stress; $\sigma_n$ is the normal stress; c is the cohesion, and $\varphi$ is the internal friction angle. The safety factor is given by the following equation:

$$FS = \frac{c}{c_r} = \frac{\tan\varphi}{\tan\varphi_r} \tag{11}$$

c: initial cohesion; $\varphi$: initial internal friction angle; $c_r$ : reduced cohesion; and $\varphi_r$ reduced internal friction angle. In the Hoek-Brown criterion, the reduced parameters in the strength reduction analysis are the initial uniaxial compressive strength of intact rock and the initial parameter of intact rock. The safety factor is given by:

$$FS = \frac{\sigma_{ci}}{(\sigma_{ci})_r} = \frac{m_i}{(m_i)_r} \tag{12}$$

$\sigma_{ci}$ the initial uniaxial compressive strength of the intact rock; and $m_i$ the parameter of the intact rock. ($\sigma$ )$_{cir}$

the reduced uniaxial compressive strength of the intact rock; and (m )$_{ir}$ the reduced intact rock parameter.

## I.8.4 Three-dimensional analysis

Usually, two-dimensional analysis is used for slope evaluation, assuming a plane deformation condition, as this ignores the third dimension of the problem. However, in nature, all slope failures are in three dimensions (3D).

Three-dimensional slope stability analysis is more realistic than 2D, but is not always used by geotechnical engineers. This is due to a number of causes and limitations of 3D stability analysis. Firstly, it is more difficult to enter data to visualise the 3D shape, [73]. Secondly, the direction of slip is not taken into account in most existing 3D MEL slope stability formulations. Thus the symmetry in geometry and loading consideration in the problems; thirdly, the existing methods in MEL are numerically unstable under transverse horizontal forces; fourthly, the location of the critical sliding surface is non-spherical; and is not yet a traditional analysis in practice to assess slope stability. However, 3D slope analysis is economical compared to 2D analysis, [74]. Popular studies have shown that the 2D safety factor is conservative for 3D slope stability analysis, [75-77]. Furthermore, the rapid development of calculation and design will make 3D slope stability analysis very accessible to geotechnical engineers. 3D slope stability analysis has been presented in several papers in the literature such as: Seed et al [78], Duncan, [28], Stark and Eid, [79], Griffiths and Marquez, [80] Michalowski, [77]. With the addition of the third dimension to 2D in the limit equilibrium method. The slices extend to the columns. Figure. 1.14 illustrates the 3D sliding surface of a slope with sliding direction and Figure, 1.15 shows inter-column forces. When W is the weight of the column, N is the normal force, T is the shear force and the resultant force between columns is Q, the width of the column is $\Delta x$ in the x direction and the length of the column is $\Delta y$ in the y direction. Furthermore, the angles of inclination of the base of the column with respect to the horizontal axes in the xz and yz planes are $\alpha_{xz}$ and $\alpha_{yz}$ respectively. The assumptions of 3D methods generally result from the essential 2D elements related to other additional conditions of the third dimension, such as the sliding direction, the inter-column forces and the 3D shape of the sliding surface. There are 3D limit equilibrium methods that ignore some of the additional definitions (forces, inter-column force, etc.) that create differences between them, and simplify some of these methods to others.

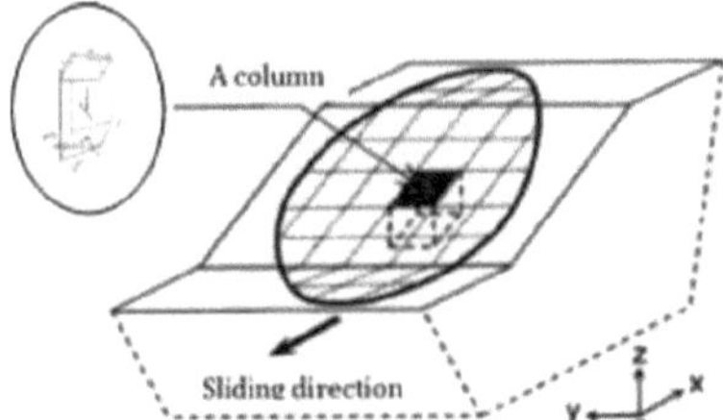


Figure. 1.14 A mass slide through vertical columns [81].

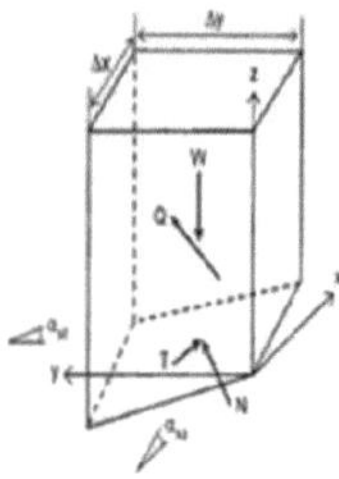

Figure. 1.15 Forces acting on a typical single column

Several 3D methods are based on 2D methods and then extended to 3D. For example, Hovland, [82], and Ugai, [83], which can be considered as the addition of the 2D method of Fellenius (1927). The extension of Bishop's (1955) 2D method to 3D methods by Ugai, [83], Hungr et al [84], Cheng and Yip, [85]. Morgenstern and Price's (1965) method was further developed in 3D by Anagnosti, 1969, [86], Cheng and Yip, [85] and Sun et al, [87]. Spencer's (1967) method has been extended to the 3D method [83,88-91]. Also, Ugai, [83], Hungr et al, [84] Yamagami and Jiang, [92, 93] Cheng and Yip, [85], present the development of the 2D Janbu (1973) method in 3D. 3D limit analysis, for slope stability is quite rare to date, Michalowski presented a 3D slope stability method for the limit analysis (upper bound) technique [94]. Li et al. produced stability diagrams for homogeneous and inhomogeneous undrained slopes in 3D by upper and lower bound analysis. Many studies based on boundary analysis have used the upper boundary analysis method alone to assess the stability of slopes, [96-100]. Recently, 3D by FEM is used for slope stability assessment [80, 101], investigates the effect of 3D on slope stability with respect to some factors, such as the geometric characteristics of the dam. Chen et al. improved the efficiency of the c-phi reduction technique to make it attractive for large-scale 3D slope stability analysis. We have also used the 3D finite element method in other studies such as Cai and Ugai, [102], Zheng et al, [103] Wei et al, 2009 [48] and Huang and Jia, [104]. Other methods used for 3D, such as finite difference and probability analysis, are used for slope stability assessments, ZHANG et al. The discrete element method was used for the evaluation of the rock slope "Acropolis Hill" in Athens by Stefanou and Vardoulakis, [106]. Figure 1.16 shows the analysis of the slope stability in 3D by the discrete element method, this example is treated by Luc and Donzé, [107]. The 3D manifold method was developed to analyse the stability of fractured rock slopes by L. He et al [108].

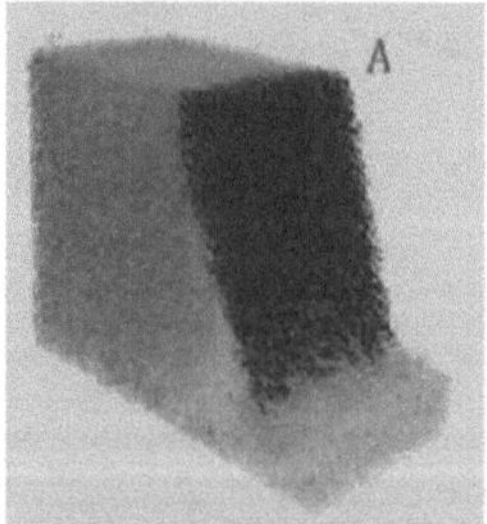
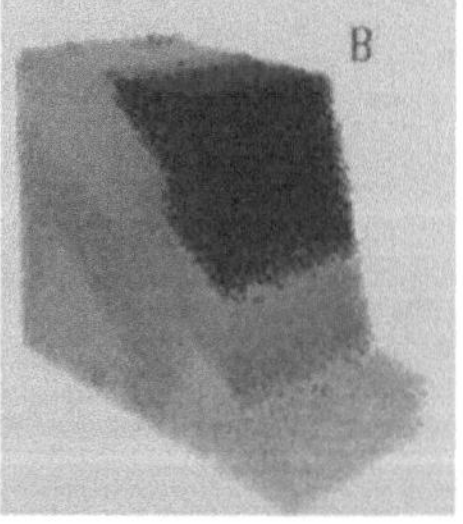


Figure. 1.16 3D slope stability analyses using the discrete element method, [107]

**PART II**

**CASE STUDY DOUNIA PARK ALGER, ALGERIA**

The objective of this chapter is to analyse a landslide "case of Parc Dounia ALGER, ALGERIA" by two different basic methods, the first one is based on the reduction of the mechanical parameters (c-phi reduction) with the help of PLAXIS 2D version 8.2 program and the second one is based on the slice method with the help of GEOSLOP software (SLOP/W).

## 11.1 Introduction

The development of the Algerian population during the last years, and its concentration in the NORTH of the country, incites the builders to project their projects on sloping grounds, the latter sometimes unstable or they become, following the intervention of the man created an imbalance (earthworks, works at the foot of the slopes...). Slope instability is one of the most damaging natural hazards in northern Algeria [109-113]. Landslides generally impose bad consequences in populated areas, they are the reason for the loss of human lives and very high costs [7,9]. In figures, the costs of landslides are between 1 and 5 billion dollars per year for each of the following countries Schuster [7]: USA, India, Japan and Italy. In Canada, the cost is about USD 70 million [8].

Geotechnical experts propose several methods for landslide assessment and prediction. These methods are presented above (Part I).

The aim of this work is to make a comparative study between two different base slip calculation methods, the first one is based on the c-phi reduction method (MEF), while the second one is based on the slice method (MEL). We have chosen as case study the slide of Parc Dounia ALGER, ALGERIA.

## 11.2 Presentation of the study area

Parc Dounia is located in the north of the capital (Algiers) exactly at Dely Ibrahim, see figure. 2.1 (red colour), near the Mercedes house. Dely Ibrahim, bordered to the North by Beni Messous, to the North-East by Bouzareah, to the East by Ben Aknoun, to the West by Cheraga, to the South by Oueld Fayat and El Achour, see figure. 2.2. The park, with a total area of 65 hectares, is bordered by the highway of the National Road No. 01, to the North the Wilaya road to the South. There are dwellings to the east and the rest of the field to the west (figure, 2.3). It is mainly concerned with a large natural park, green areas, recreational areas and educational facilities.

Figure. 2.1 Position of Dely Ibrahim (Google map)

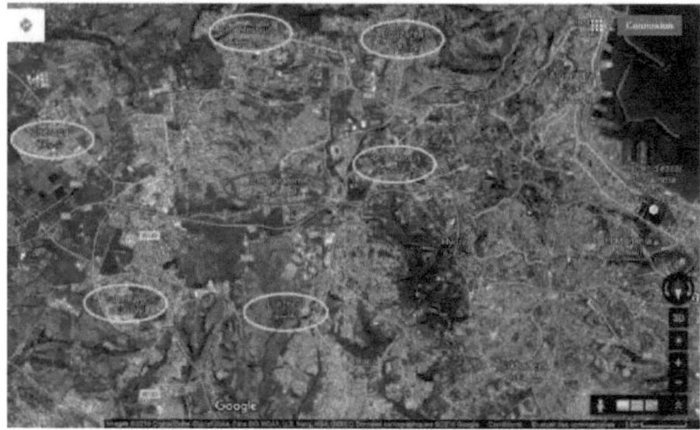

Figure. 2.2 Google map of the location of Dely Ibrahim and its borders (Google map)

Figure. 2.3 Position of Dounia Park and its boundaries

Before analysing a landslide, several indispensable steps such as (a geological survey, the state of the area, geotechnical reconnaissance, geometry and soil properties), all these steps were done by the laboratory of habitat and construction LHC unit Ain Taya, Algiers, Algeria.

11.2.1 Geological survey

It is impossible to make a landslide analysis without a geological survey, the latter aims to get an idea of the nature of the terrain, the nature of the bedrock and the presence or absence of faults.

A geological map of the area, drawn up in 1960, shows the following geological formations (Figure 2.4). This area is sensitive to water, as it is largely made up of marl formations. Altered marl has a high porosity due to its reworking and loses its cohesion in the presence of water. Depending on its degree of alteration, the plaisancian marl has different behaviours with regard to the stability of the slope and its resistance to breaking.

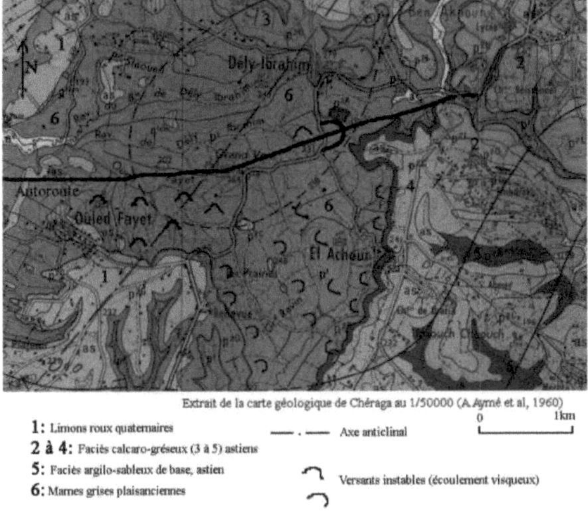


Figure. 2.4 Geological formations of the Cheraga map (at 1 / 50,000)

## 11.2.2 The state of the region

This step allows us to make a thorough geological study, which is based on :

- Limiting the slip zone;
- Efficient implementation of geotechnical tests (core drillings, boreholes
  (pressure meters, inclinometers, etc.) and geophysical (seismic profiles, electrical imaging, etc.);
- Examination of photographs (the shape of trees, cracks in buildings and open cracks in roads and
  paths...).

After the site visit, Figure 2.5 shows the landslide area. This area is limited by the yellow colour.
Several observations of the terrain and its surroundings (Dounia Park) have revealed the existence of
instabilities, such as

- A slope with a significant angle that makes slopes easily unstable in combination with other factors
  (water, earthquake) figure 2.6 a ;
- The shape of the trees is tilted towards the slope as shown in Figure 2.6 b;
- Due to the nature of the soil, the top layer (fill) is slippery on the marl layer, due to the large thickness
  of the fill (enter 2 to 3 m) with the high rainfall of this area (figure 2.7);
- Open cracks in the ground (Figure 2.8);
- Significant discontinuity on the road (Figure 2.9)

Figure. 2.5 Area of geotechnical investigations bounded by the colour yellow

Figure. 2.6 (a) The slope, (b) The inclination of the trees

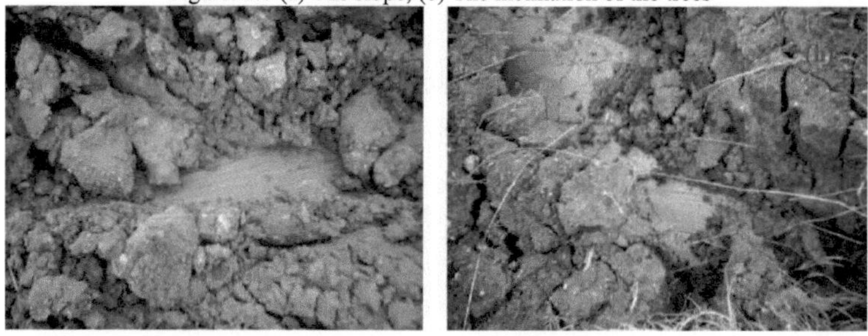

Figure. 2.7 Observations indicating soil instability

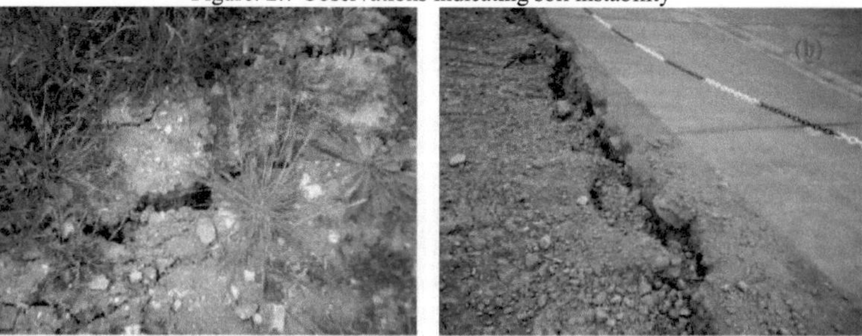

Figure. 2.8 Cracks observed in the ground and near the road N: 01

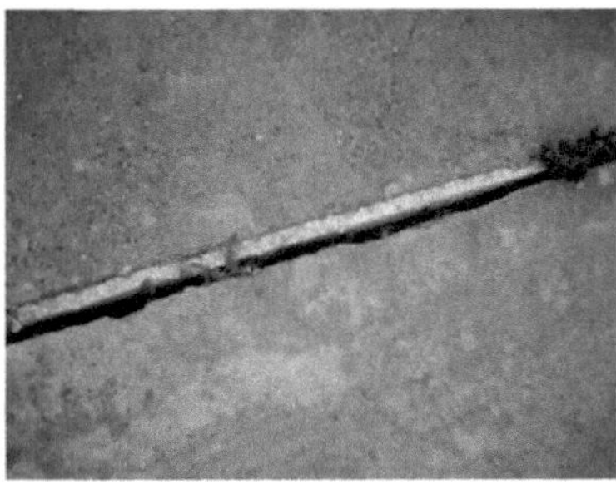

Figure. 2.9 Fissure observed in and around the
Dounia Park area

This area is affected by many types of landslides. Due to the morphological characteristics of landslides, and according to the classification of Cruden and Varnes [2]: translational and rotational landslides, as seen in figures 2.10 and 2.11 Translational landslides, of very variable size, affected by marl and fill, and the angle of the slope (figure. 2.10 (a) and (b)). The second type of landslide is rotational and is illustrated in Figure 2.11 (a) and (b). These landslides have different sizes and precipitation plays an important role in this instability. Translational landslides are gentler slopes compared to rotational landslides.

Figure. 2.10 Terrain instabilities, through translational slides

Figure. 2.11 Two rotational slides

## 11.2.3 Geotechnical investigations

The geotechnical study was carried out by the Laboratory of Housing and Construction (LHCC 2011) unit Ain Taya (Algiers, ALGERIA). The investigation and reconnaissance campaign on the site involved the performance of in situ tests (core drillings, wells, pressuremeter tests, dynamic penetrometer) and laboratory tests (physical identification, shear tests, odometer tests). The purpose of this reconnaissance is to :

- Determine the soil lithology;
- Determine the physical and mechanical characteristics of the different soil layers;
- To have an idea of the degree of soil homogeneity and dynamic resistance as a function of depth;
- Detecting the water level ;
- Determine the site classification (limiting pressure and pressure modulus)

According to the lithological cross-section of the core drillings, wells and auger drillings, the soil is essentially made up of a layer of compacted marl (plaisancian marl) which is considered to be the substratum, surmounted by altered marl of thickness (approximately 8 m) and a significant layer of fill of variable thickness (0.2 to 4 m).

The water level is detected by piezometer measurements at variable levels (0.5 to 10 m), so the formations are saturated with water

The geotechnical characteristics are presented in Table 2.1

Table 2.1: Geotechnical characteristics

| Parameters | Designations | Backfill | Marly clay (weathered marl) | Compacted marl |
|---|---|---|---|---|
| Pattern and type of behaviour | | M.C. | M.C. | M.C. |
| Bulk density | yh (kN/m )$^3$ | 14.5 | 19.60 | 19.50 |
| Saturated density | ys (kN/m )$^3$ | 16.50 | 22.40 | 21.90 |
| Poisson's ratio | $\upsilon$ | 0.499 | 0.300 | 0.300 |
| Young's modulus | E (KN/m )$^2$ | $2.10^4$ | $10^5$ | 106 |
| Cohesion | C (KN/m )$^2$ | 05 | 27 | 40 |
| Friction angle | $\varphi$ (°) | 15 | 14 | 14 |
| Expansion angle | $\Psi$ (°) | 0 | 0 | 0 |

## 11.2.4 Soil geometry and properties

Based on the topographic survey provided by the LHCC laboratory, we plotted the shape of the area in 3D using SURFER software (Figure 2.12 (a)) [114]. We chose the most critical slope (Figure 2.12 (b)) for the analysis, the length of which is 237 m and the height is 59 m.

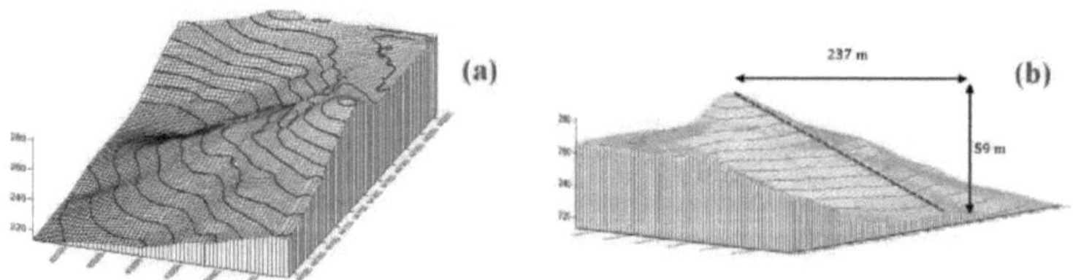

Figure. 2.12 (a) The study area in 3D, (b) The choice of the studied profile

## 11.3  Numerical modelling and evaluation

The slope is simulated under a two-dimensional plane deformation assumption, the Mohr-Coulomb model, and all soil property characteristics are presented in the previous table (Table 2.1).

### 11.3.1  Stability analysis by limit equilibrium methods

After the evaluation of the slope with the finite element method, we analysed the same slope with the limit equilibrium methods, to make a comparison between the results of the sliding surfaces and the safety factors. The stability study was carried out using GEOSTUDIO software (SLOP/W module). Figures 2.13, 2.14, 2.15 and 2.16 show the slip surfaces obtained by the ordinary method, the BISHOP method, the JANBU method and the Morgenstern price method respectively. The corresponding safety factor values are presented in Table 2.2 :

Table 2.2: Slope safety factor

| Method | Value of Fs |
|---|---|
| Ordinary | 0.810 |
| BISHOP | 0.910 |
| JANBU | 0.812 |
| Morgenstern Price | 0.911 |

Figure 2.13 shows the sliding surface with the limit equilibrium method (Ordinary method)

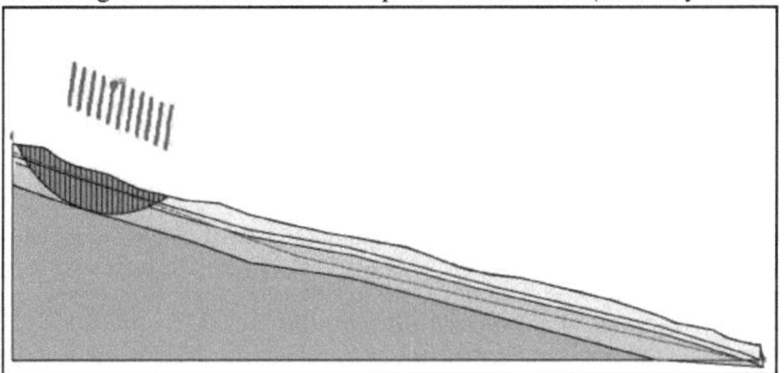

Figure. 2.13 Slip surface obtained by the Ordinary method

The second method used to evaluate the sliding surface of our slope is the Bishop method (Figure 2.14).

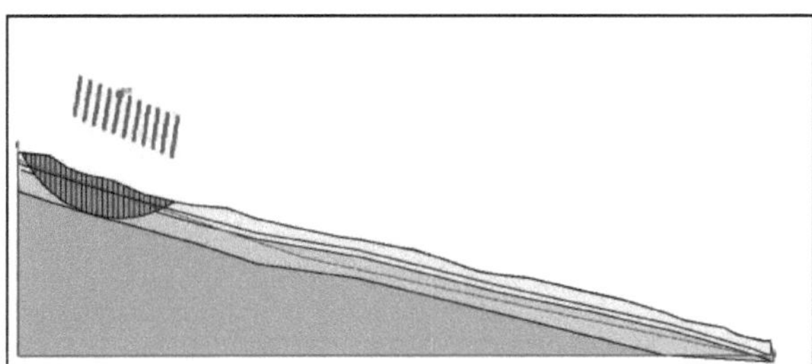

Figure. 2.14 Slip surface obtained by Bishop's method

The third method used is the Janbu method (Figure 2.15)

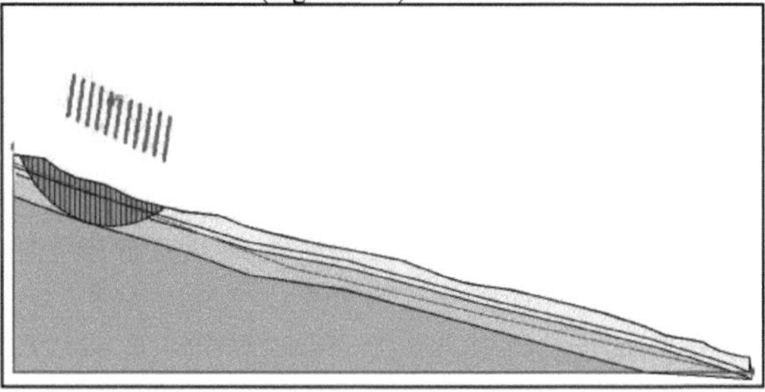

Figure. 2.15 Slip surface obtained by the Janbu method

The last method used to evaluate the sliding surface is the Morgenstern Price method (Figure 2.16).

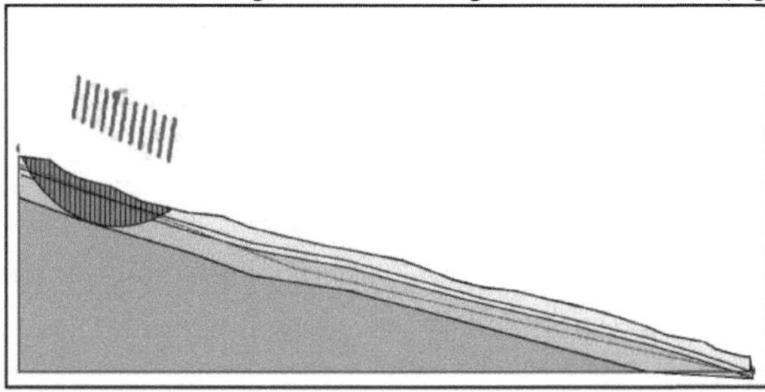

Figure. 2.16 Slip surface obtained by the Morgenstern Price method

The previous results show that the different slice methods (ordinary, Bishop, Janbu, Morgenstern and price) give the same position of the slip zone, and a safety coefficient is close between them.

## II.3.2 Numerical modelling using the finite element method

The c-phi reduction method was chosen to analyse this landslide, using PLAXIS 2D version 8.2 software.

This technique is based on the progressive reduction of mechanical parameters (C, $\varphi$) until failure.

The reference model is made by triangular elements with 15 nodes, the number of elements is 226 and the number of nodes is 1953.

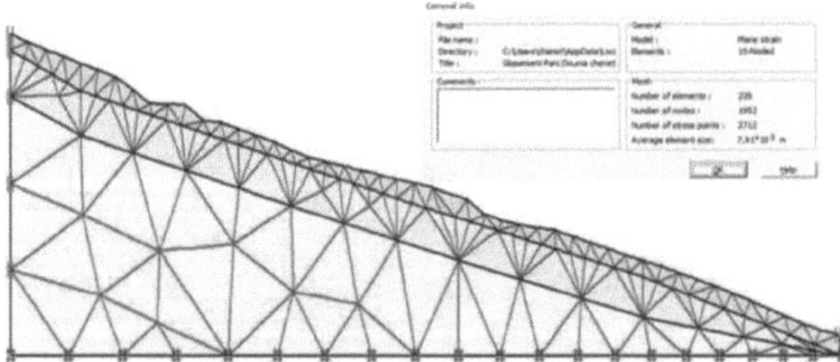

Figure. 2.17 Mesh and geometric boundary conditions of PLAXIS model

The safety coefficient is given by equation (11 part I), the analysis is carried out in two steps:

> **Phase 0:** Initiation of constraints ;

> **Phase 1:** Calculation of safety coefficient.

$$FS = \frac{c}{c_r} = \frac{\tan \varphi}{\tan \varphi_r}$$

The surface was automatically detected with the lowest safety factor (figure 2.18), we obtained a safety factor Fs = 1.059

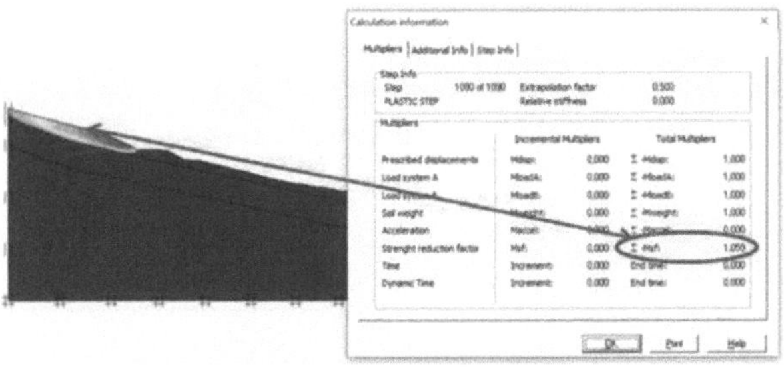

Figure. 2.18 Safety coefficient calculated by PLAXIS

Figure 2.19 shows the failure surface of our slope, from the shape of this surface we can classify this type of slip as a rotational slip (Figure 2.19).

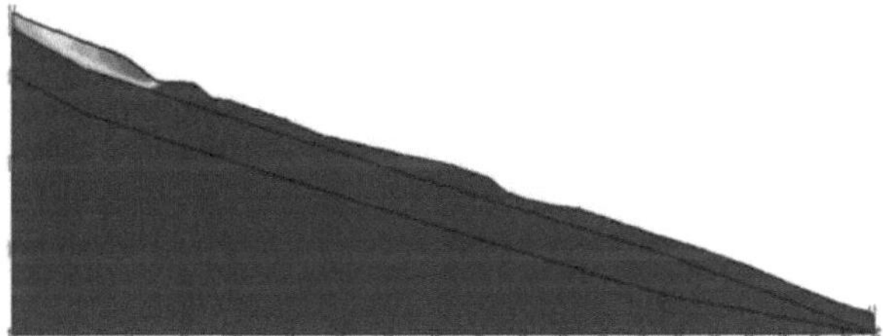

Figure. 2.19 The critical sliding surface

Figure 2.20 shows that the slip surface is just in the top layer of the soil (fill), this was pointed out earlier during our site visit, this confirms that the finite modelling is close to reality.

Figure 2.21 shows the total break length Ltr and the break length Lr.

The two lengths (Ltr=48.50 m and Lr=33 m) can be calculated on the basis of the sliding surface.

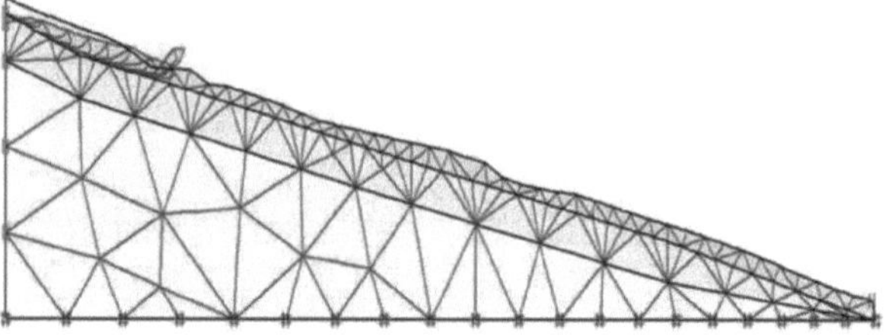

Figure. 2.20 The fracture surface by mesh deformation

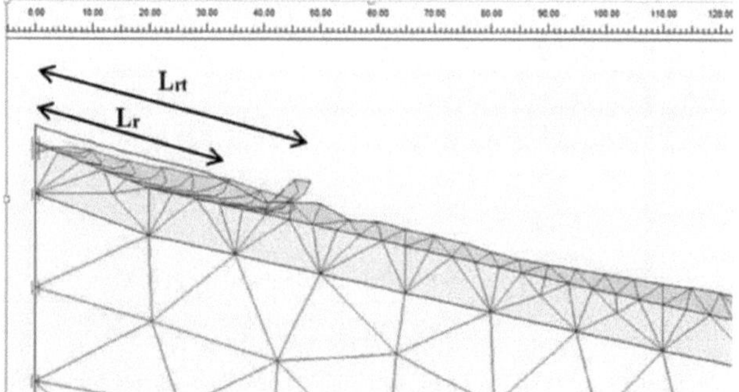


Figure. 2.21 Zooming in on the fracture surface by mesh deformation

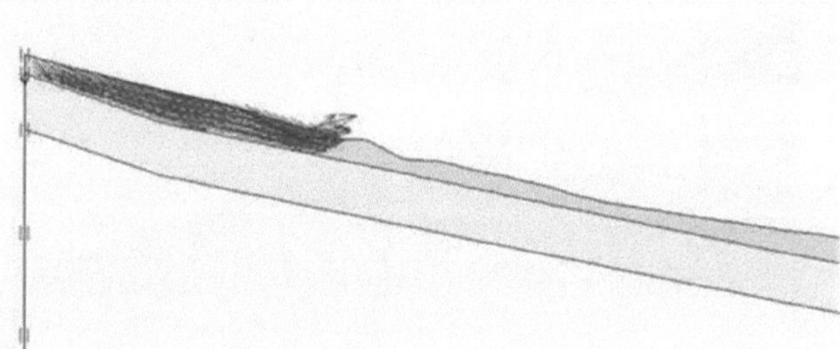

Figure. 2.22 The displacement field rupture surface

Figures 2.21 and 2.22 show the slope displacement field. In these figures we can clearly see that the upper fill layer slides over the marl layer due to the high fill thickness, the high water level and the low permeability of the marl.

## II.4 Conclusion

After modelling by the limit equilibrium method and the finite element method we noticed that :

- the safety factors are small by both methods (< 1).
- The depth of sliding by the MEL is large compared to the MEF, it affects the second layer (Marly clay);
- The position of the sliding surface is the same for both methods.

## GENERAL CONCLUSION

The objective of this work is to analyse a landslide on the one hand; to determine the landslide area and the corresponding value of the safety coefficient and on the other hand, a comparative study on the analysis of a landslide by the finite element method and the limit equilibrium method.

Based on the analysis of the slip results by both PLAXIS 2D and SLOP/W software, it was found that :

- The standard deviation between the safety factor results by the limit equilibrium methods is 0.11.
- Landslide modelling by two different software packages gives close results, the same position of the landslide area and the safety factor for the different methods (MEL, MEF) is less or equal to 1.
- The safety factor value obtained by the FEM is a 10% difference from the value obtained by the Bishop method. This difference is judged by the inaccuracy of the measurement of the mechanical resistance characteristics (C, ^).
- The finite element method is a very important method to analyse a landslide, because it determines: the critical slide area, the lowest safety factor value, the stresses and strains in a soil mass, and even can easily vary the geotechnical, hydraulic, boundary conditions and the material behaviour law
- The finite element analysis of the landslide is close to reality, as the landslide surface is located in one lithological layer (fill layer). However, the landslide surface by the MEL is cut into the second layer (marly clay).

# REFERENCE

[1] Hamza-Cherif Riad, Magister thesis in Civil Engineering. Theme: Etude Des Mouvements De Pentes Par Le Code De Calcul "Pfc2d". Université Abou-Bekr Belkaid (Tlemcen). 2009.

[2] Cruden, D.M., Varnes, D.J. Landslide type and processes". In: Turner, A.K., Schuster, R.L. (Eds.), Landslides. Investigation and Mitigation. Special Report 247. Transportation Research Board. National Academy Press, Washington, pp. 36-75. 1996.

[3] Varnes, D. J. Slope movement types and processes. In Schuster, R. L. and Krizek, R. J., editors, Landlides: analysis and control, volume 176, pages 11{33. National Academic Press, Washington, USA. 1978.

[4] Cornforth, D. H. Landslides in practice: investigation analysis, and remedial/preventive option in soils. Wiley and Sons, Hoboken, USA. ISBN 0-471-67816-3. 2005.

[5] Kim, J.Y., Lee, S.R. An improved search strategy for the critical slip surface using finite element stress fields. Computers and Geotechnics 21, 295-313. 1997.

[6] Schuster, RL. Socioeconomic significance of landslides. Landslides - Investigation and Mitigation, Special Report 247, A.K. Turner & RL. Schuster (Eds.) Transportation Research Board, Washington, DC, pp. 12-35. 1996.

[7] Sidle CR, Ochiai H. Landslides. Processes, prediction, and land use. AGU Books, Washington, DC. 2006.

[8] Brabb, E.E. & Harrod, B.L. (Eds), Landslides: Extent and Economic Significance, In Proceedings, 28th International Geological Congress: Symposium on Landslides, Washington, DC. 1989.

[9] Djamel Eddine Benouis. Study of a landslide by different methods. Thesis of magister. University of Saida. 2005.

[10] Ilhem Tir. Consequence of the Constantine landslides. Le Soir. 2003.

[11] Taleb Hosni Abderrahmane, Berga Abdelmadjid, book. Review of slope stability assessment. Publisher: LAP LAMBERT Academic publishing, 2017.

[12] Sujit Mandal. Ramkrishna Maiti. Semi-quantitative Approaches for Landslide Assessment and Prediction. Springer Natural Hazards. 2015.

[13] Fabio Vittorio De Blasio. Introduction to the Physics of Landslides, Lecture Notes on the Dynamics of Mass Wasting. Springer. 2011.

[14] Williams, G.P., and Guy, H.P. Erosional and depositional aspects of Hurricane Camille in Virginia, 1969: U.S. Geological Survey Professional Paper 804, 80 p. 1973.

[15] Jibson, R.W. Evaluation of landslide hazards resulting from the October 5-8, 1995 storm in Puerto Rico: U.S. Geological Survey Open-File Report 86-26, 40 p. 1986.

[16] Lashkaripour. G,R. Hazard evaluation of landslide in Iran (1999), slope stability engineering Volume 2; Proceedings of the International Symposium on Slope Stability Engineering - Is-Shikoku'99/Matsuyama/Shikoku/Japan1 /1 N8-O Vember. 1999.

[17] Graziella Devoli. Collection of Data on Historical Landslides in Nicaragua, Chapter 29, Landslides Risk Analysis and Sustainable Disaster Management, Proceedings of the First General Assembly of the International Consortium on Landslides, Springer-Verlag Berlin Heidelberg 2005.

[18] Haugen, E.D. and Kaynia, A.M. Vulnerability of structures impacted by debris flow; Landslides and Engineered Slopes - Chen et al. (eds) 2008 Taylor & Francis Group, London, ISBN 978-0-415-41196-7. 2008.

[19] Hoe I. Ling, Dov Leshchinsky, Nelson N.S. Chou. Post-earthquake investigation on several geosynthetic-reinforced soil retaining walls and slopes during the Ji-Ji earthquake of Taiwan. 2001.

[20] Crosta, G.B,T, Imposimato. S, Roddeman. D, Chiesa. S, Moia. F. Small fast-moving f low- like landslides in volcanic deposits: The 2001 Las Colinas Landslide (El Salvador); Engineering Geology 79 185-214. 2005.

[21] http://en.wikipedia.org/wiki/List_of_landslides

[22] Michael Duncan, J. Stephen G. Wright. Soil Strength and Slope Stability. JOHN WILEY & SONS, INC. 2005.

[23] Seed, H. B. Landslides During Earthquakes due to Soil Liquefaction. Terzaghi Lecture, Journal of the

Soil Mechanics and Foundations Division, ASCE. Vol. 92, No. SMI. pp. 1341. 1968.

[24] Lee W. Abramson, Thomas S. Lee, Sunilsharma, Glenn M. Boyce. Slope Stability And Stabilization Methods (Second Edition). John Wiley & Sons, Inc. 2002.

[25] Blylh. F.G. R.and M. H. de Freilas. A Geology for Engineers. 7th ed. New York: Elsevier. 1984.

[26] Bjerrum. L. Progressive Failure in Slope in Overconsolidated Plastic Clays and Clay- Shales. Terzaghi Lecture. Journal of Soil Mechanics Foundation Division, ASCE, Vol. 93. No. SMS. pp. 3-49. 1967.

[27] Vaunat J., Leroueils, Faure R.M. Slope movements: a geotechnical perspective, 7th IAEG congress, Lisbon, pp. 1637-1646, 1994.

[28] Duncan J M. State of the Art: Limit Equilibrium and Finite-Element Analysis of Slopes. Journal of Geotechnical Engineering. 122:577-596. 1996.

[29] Haefeli R. The Stability of Slopes Acted Upon by Parallel Seepage, Proceedings of the Second International Conference on Soil Mechanics and Foundation Engineering, Rotterdam. 1948;Vol. 1, pp. 57-62.

[30] Taylor. D. W. Stability of Earth Slopes. Journal of the Boston Society of Civil Engineers. 1937; Vol. 24, pp. 197-246.

[31] LEE W. ABRAMSON, THOMAS S. LEE, SUNILSHARMA, GLENN M. BOYCE. Slope Stability and Stabilization Methods (Second Edition). JOHN WILEY & SONS, INC; 2002.

[32] J Michael Duncan, Stephen G. Wright. Soil Strength and Slope Stability. JOHN WILEY & SONS, INC; 2005.

[33] Y.M. Cheng and C.K. Lau, Slope Stability Analysis and Stabilization New methods and insight. This edition published in the Taylor & Francis e-Library; 2008.

[34] Fellenius W. Calculation of stability of earth dams. Transactions, 2nd Congress on Large Dams, Washington, DC. 1936; vol1445-462

[35] Bishop A.W. The use of slip circle in the stability analysis of slopes. Geotechnique. 1955; 5(1):7-17.

[36] Morgenstern N. R., and Price V. E. The analysis of the stability of general slip surfaces. Geotechnique, London. 1965; 15(1), 79-93.

[37] Spencer E. A method of analysis of the stability of embankments assuming parallel interslice forces. Geotechnique. London. 1967; 17(1), 11-26.

[38] Janbu N. Slope stability computations. Soil Mech. and Found. Engrg. Rep. The Technical University of Norway, Irondheim, Norway; 1968.

[39] Lowe J, and Karafiath L. Stability of earth dams upon drawdown. Proc. 1st Pan-Am. Conf. on Soil Mech. and Found. Engrg, Mexico City. 1960; 2, 537-552.

[40] U.S. Army Corps of Engineers. Engineering and design-stability of earth and rock fill dams. Engr. Manual EM 1110^2-1902, Dept. of the Army, Corp of Engrs. Ofc. of the Chf. of Engrs. 1970.

[41] Sarma S K. Stability Analysis of Embankment and Slopes. Geotechnique. 1973; 23:423033.

[42] *Fredlund, D. G., and Krahn, J. Comparison of slope stability methods of analysis. Can. Geotech. J. 1977; 14(3),429-439.*

[43] *Transportation Research Board. Landslides: Investigation and Mitigation. TRB Special Report 247. National Academy Press, Washington, D.C., U.S.A. 1996.*

[44] *Vidar Thomée. From finite differences to finite elements a short history of numerical analysis of partial differential equations. Journal of Computational and Applied Mathematics. 2001; 128 1-54.*

[45] *Antonio Bobet. Numerical Methods in Geomechanics. The Arabian Journal for Science and Engineering, Volume 35, Number 1B. 2010.*

[46] *ITASCA. FLAC2D (Fast Lagrangian Analysis of Continua) version 5.0 user's manual. Itasca Consulting Group, Minneapolis, MN, USA. 2005.*

[47] *Dawson EM, Roth WH, Drescher A. Slope stability analysis by strength reduction. Geotechnics 1999; 49(6):835-40.*

[48] *Wei, W.B., Cheng, Y.M. Soil nailed slope by strength reduction and limit equilibrium methods. Computers and Geotechnics. 2010; 37 (5), 602e618.*

[49] *Xinpo Li, Siming He, Yu Luo and Yong Wu. Numerical studies of the position of piles in slope stabilization. Geomechanics and Geoengineering: An International Journal. 2011; 6:3, 209-215.*

[50]    *Iman Mehdipour, Mahmoud Ghazavi, Reza Ziaie Moayed. Numerical study on stability analysis of geocell reinforced slopes by considering the bending effect; journal of Geotextiles and Geomembranes. 2013.*

[51]    *E Hinton, D R J. Owen. Finite Elements in Plasticity: Theory and Practice, Pineridge Press, Swansea, Wales, 1980.*

[52]    *O C Zienkiewicz, R L Taylor. 4th edition, The Finite Element Method, vol. 2, McGrawHill, London, 1991.*

[53]    *M.A. Crisfield, (1991). Non-linear Finite Element Analysis of Solids and Structures, John Wiley, Chichester,*

[54]    *K.J. Bathe (1996). Finite Element Procedures, Prentice-Hall, New Jersey.*

[55]    *Colby, C. Swan and Young-Kyo Seo. 1999 Slope stability analysis using finite element techniques. 13 th Iowa ASCE Geotechnique Conference, USA.*

[56]    *L.C. Li, C.A. Tang, W.C. Zhu, Z.Z. Liang. Numerical analysis of slope stability based on the gravity increase method, journal of Computers and Geotechnics. 2009.*

[57]    *Tang CA. Numerical simulation on progressive failure leading to collapse and associated seismicity. Int J Rock Mech Min Sci. 1997; 34(2):249-61.*

[58]    *Zienkiewicz OC, Humpheson C, Lewis RW. Associated and nonassociated viscoplasticity and plasticity in soil mechanics. Géotechnique 1975;25(4):671-89.*

[59]    *Smith IM, Griffiths DV. Programming the finite element method. 2nd ed. John Wiley & Sons; 1982.*

[60]    *Naylor DJ. Finite elements and slope stability. Proceedings of the NATO Advanced Study Institute, Lisbon, Portugal, 1981. Numer. Methods Geomech. 1982:229-44.*

[61]    *Donald IB, Giam SK. Application of the nodal displacement method to slope stability analysis. In: Proceedings of the 5th Australia New Zealand conference on geomechanics, Sydney, Australia, 1988. p. 456-60.*

[62]    Ugai K. A method of calculation of total factor of safety of slopes by elastoplastic FEM. Soils Foundations 1989;29(2):190-5.

[63]    Matsui T, San KC. Finite element slope stability analysis by shear strength reduction technique. Soils Found 1992;32(1):59-70.

[64]    Ugai K, Leshchinsky D. Three-dimensional limit equilibrium and finite element analysis: a comparison of results. Soils Found 1995;35(4):1-7.

[65]    Song E. Finite element analysis of safety factor for soil structures. Chinese J Geotech Eng 1997;19(2):1-7.

[66]    Griffiths DV, Lane PA. Slope stability analysis by finite elements. Géotechnique.1999;49(3):387-403.

[67]    Hammah RE, Yacoub TE, Corkum B, Curran JH. A Comparison of finite element slope stability analysis with conventional limit-equilibrium investigation. In: Proceedings of the 58th Canadian geotechnical and 6th joint IAH-CNC and CGS groundwater specialty conferences Saskatoon, Saskatchewan, Canada, September; 2005.

[68]    Zheng YR, Zhao SY, Kong WX, Deng CJ. Geotechnical engineering limit analysis using finite element method. Rock Soil Mech 2005;26(1):163-8.

[69]    Taleb H A, Berga A. Parametric Study of the Coupling (C, phi) and its Influence on the Slip Safety Factor by the Limit Equilibrium Method. 1st International Seminar on Civil Engineering, University of Bechar. 2013.

[70]    Hammah R E, Yacoub T E, and Corkum B C. The Shear Strength Reduction Method for the Generalized Hoek-Brown Criterion, ARMA/USRMS 05-810. 2005.

[71]    Taleb Hosni Abderrahmane, Berga Abdelmadjid. Numerical analyses of slope stability using the Hoek-Brown and Mohr-Coulomb criterions. The second international seminar of Civil Engineering 'SIGCB'.,2015.

[72]    Hoek E, C Carranza-Torres, and B Corkum. Hoek-Brown criterion □ 2002 edition. In Proceedings of the 5th North American Rock Mechanics Symposium and the 17th Tunnelling Association of Canada: NARMS-TAC, Toronto, Canada, eds. R.E. Hammah et al, 2002;Vol. 1, pp. 267-273.

[73]    Leong E C, H Rahardjo. Two and three-dimensional slope stability reanalyses of Bukit Batok slope

2012; Computers and Geotechnics.

[74] Cavounidis S. On the ratio of factors of safety in slope stability analyses. Geotechnique 1987;37 (2), 207-220.

[75] Li AJ, Merifield RS, Lyamin AV. Limit analysis solutions for three dimensional undrained slopes. Comput Geotech 2009;36:1330-51.

[76] Li AJ, Merifield RS, Lyamin AV. Three-dimensional stability charts for slopes based on limit analysis methods. Can Geotech J 2010;47:1316-34.

[77] Michalowski RL. Limit analysis and stability charts for 3D slope failures. J Geotech Geoenviron Eng 2010;136(4):583-93.

[78] Seed R B, Mitchell J K, and Seed, H.B. Kettleman Hills waste landfill slope failure. II Stability Analysis. J Geotech Eng ASCE 1990;116(4): 669-690.

[79] Stark T D, and Eid H T, Performance of three-dimensional slope stability methods in practice." J Geotech Geoenv Eng, 1998;124 (11): 1049-1060.

[80] Griffiths D V, and Marquez R.M. Three-dimensional slope stability analysis by elasto- plastic finite elements. Geotechnics, 2007;57(6): pp.537-546.

[81] Hajiazizi M., Tavana H. Determining three-dimensional non-spherical critical slip surface in earth slopes using an optimization method. Engineering Geology 2013;153 114-124.

[82] Hovland H. Three-dimensional Slope Stability Analysis Method. Journal of Geotechal Engineering Division 1977;103(9),971-986.

[83] Ugai K. Three-dimensional Slope Stability Analysis by Slice Methods. Proceeding of the International Conference on Numerical Methods in Geomechanics. Innsbruck, Austria 1988;1369-1374.

[84] Hungr O, Salgado F, and P Byrne. Evaluation of a three-dimensional method of slope stability analysis. Canadian Geotechnical Journal 1989;26, 679-686.

[85] Cheng Y, and C Yip. Three-Dimensional Asymmetrical Sloe Stability Analysis Extension of Bishop's, Junbu's, and Morgenstern-Price's Techniques. Journal of Geotechnical and Geoenvironmental Engineering, 2007;133 (12), 1544-1555.

[86] Anagnosti P. Three dimensional stability of fill dams," Proceeding of 7th International Conference on Soil Mechanics and Foundation Engineering, Mexico 1969;275-280.

[87] Sun G, Zheng H., and Jiang W. A global Procedure for Evaluating Stability of Three- Dimensional Slopes. Natural Hazards 2012; 61(3): 1083-1098.

[88] Chen R, and J Chameau. Discussion Three-dimensional Limit Equilibrium Analysis of Slopes. Geotechnique 1983 b;33(1), 215-216.

[89] Xing Z. Closure: Three-Dimensional Stability Analysis of Concave Slopes in Plan View. Journal of Geotechnical Engineering, 1988b;116(2), 345-346.

[90] Chen Z, Mi H., Zhang F, and X Wang. A Simplified Method for 3D Slope Stability Analysis. Canadian Geotechnical Journal 2003;40, 675-683.

[91] Jiang, J. C. and T. Yamagami Three-dimensional Slope Stability Analysis using an Extended Spencer Method. Journal of the Japanese Geotechnical Society of Soils and Foundations 2004;44(4), 127-135.

[92] Yamagami T, and J C Jiang. Determination of the sliding direction in three dimensional slope stability analysis. Proceedings of the 2nd International Conference on Soft Soil Engineering, Part 1, Nanjing: Hohai University Press 1996;567-572.

[93] Yamagami T, and J C Jiang A Search for the Critical Slip Surface in Three-dimensional Slope Stability analysis Soils and Foundations 1997;37(3), 1-16.

[94] Michalowski R.L. Three-dimensional analysis of locally loaded slopes, Geotechnique 1989;39(1), 27-38.

[95] A J Li, R S Merifield, A V Lyamin. Limit analysis solutions for three dimensional undrained slopes. Computers and Geotechnics 2009.

[96] Chen Z, Wang X, Haberfield C, Yin J-H, Wang Y. A three-dimensional slope stability analysis method using the upper bound theorem part I: theory and methods. Int J Rock Mech Min Sci 2001;38:369-78.

[97] Farzaneh O, Askari F. Three-dimensional analysis of nonhomogeneous slopes. J Geotech Geoenviron Eng ASCE 2003;129(2):137-45.

[98] De Buhan P, Garnier D. Three dimensional bearing capacity analysis of a foundation near a slope.

Soils Found 1998;38(3):153-63.

[99] Michalowski RL. Stability charts for uniform slopes. J Geotech Geoenviron Eng ASCE 2002;128(4):351-5.

[100] Viratjandr C, Michalowski RL. Limit analysis of submerged slopes subjected to water drawdown. Can Geotech J 2006;43:802-14.

[101] Yu Y, Xie L, and Zhang B. Stability of earth-rockfill dams: Influence of geometry on the three-dimensional effect. Computers and Geotechnics, 2005;32: 326-339.

[102] Cai F, Ugai K. Numerical analysis of the stability of a slope reinforced with piles. Soils Found 2000;40(1):73-84.

[103] Zheng H, Liu DF, Li CG. Slope stability analysis based on elasto-plastic finite element method. Int J Numer Meth Eng 2005;64(14):1871-88.

[104] Huang M, Jia CQ. Strength reduction FEM in stability analysis of soil slopes subjected to transient unsaturated seepage. Comput Geotech 2009;36(1-2):93-101.

[105] ZHANG Ke, CAO Ping, LIU Zi-yao, HU Hui-hua, GONG Dao-ping. 2011. Simulation analysis on three-dimensional slope failure under different conditions Trans. Nonferrous Met. Soc. China 21(2011) 2490-2502.

[106] Stefanou, I., and I. Vardoulakis (2005) -Stability assessment of SE/E rock corner slope of the Acropolis Hill in Athens,Il 5th GRACM International Congress on Computational Mechanics.

[107] Luc Scholtès, Frédéric-Victor Donzé (2012), Modelling progressive failure in fractured rock masses using a 3D discrete element method, International Journal of Rock Mechanics & Mining Sciences.

[108] L. He, X.M.An , G.W.Ma , Z.Y.Zhao. 2013 Development of three-dimensional numerical manifold method for jointed rock slope stability analysis International Journal of Rock Mechanics & Mining Sciences.

[109] Bouhadad, Y., Benhammouche, A., Bourenane, H., Ait Ouali, A., Chikh, M., Guessoum, N. (2010), "The Laalam (Algeria) damaging landslide triggered by a moderate earthquake (Mw = 5.2)", Nat Hazards, 54:261-272.

[110] Djerbal, L., Melbouci, B. (2012), "Le glissement de terrain d'Ain El Hammam (Algérie): causes et evolution" - Bulletin of Engineering Geology and the Environment, vol 71, pp 587-597.

[111] Riheb, H., Abderrahmane, B., Yacine, L., Mustapha, B., Abd El Madjid, C.c., Abdeslem, D. (2012), "Geologic, topographic and climatic controls in landslide hazard assessment using GIS modeling: a case study of Souk Ahras region", NE Algeria. Quat Int 302(2013):224-237.

[112] Guettouche, M.S. (2012), "Modeling and risk assessment of landslides using fuzzy logic. Application on the slopes of the Algerian Tell (Algeria)", Arabian Geosci, 39:1866-751. doi:10.1007/s12517-012-0607-5.

[113] Bouhadad, Y., (2013), "Occurrence and impact of characteristic earthquakes in northern Algeria", J Nat hazards, 67:1573-0840. doi:10. 1007/s11069-013-0704-0.

[114] Taleb H A. Master thesis. Land stability study with comparison between Plaxis and GEO-SLOPE. University of Zaine Achor of Djelfa. 2011.

# I want morebooks!

Buy your books fast and straightforward online - at one of world's fastest growing online book stores! Environmentally sound due to Print-on-Demand technologies.

Buy your books online at
**www.morebooks.shop**

Kaufen Sie Ihre Bücher schnell und unkompliziert online – auf einer der am schnellsten wachsenden Buchhandelsplattformen weltweit! Dank Print-On-Demand umwelt- und ressourcenschonend produziert.

Bücher schneller online kaufen
**www.morebooks.shop**

info@omniscriptum.com
www.omniscriptum.com